Introducing ecology

Nature at work

British Museum (Natural History)
Cambridge University Press

Published by the British Museum (Natural History), London,
and the Syndics of the Cambridge University Press,
The Pitt Building, Trumpington Street, Cambridge CB2 1RP
Bentley House, 200 Euston Road, London NW1 2BD
32 East 57th Street, New York, NY10022, USA
296 Beaconsfield Parade, Middle Park, Melbourne 3206, Australia

ISBN 0 521 22390 3 hard covers
ISBN 0 521 29469 X paperback

First published 1978
Reprinted 1982

Printed in Great Britain by
Balding + Mansell Limited, London and Wisbech

Contents

Preface

The world is teeming with life. Plants and animals are constantly interacting with each other and with their non-living surroundings to form an intricate network of activity.

How can we begin to understand the way such a complex system works?

We can start by looking at how energy flows through the system. The energy that all living things need comes originally from the sun. It is captured by plants and transferred to other living things as they feed.

In this book you can follow the flow of energy through an oak woodland and a rocky seashore. As you begin to appreciate the interdependence of living things and understand something of the workings of these complex systems, we hope you will discover a new way of looking at your natural surroundings.

This book has been produced in conjunction with the exhibition **Introducing ecology** which opened at the British Museum (Natural History) in October 1978. The exhibition is the second in the Museum's major new exhibition programme and represents an exciting new approach to learning about natural history.

Preparing the exhibition involved the effort and imagination of many people, both within the Museum and outside, and I should like to take this opportunity of thanking everyone concerned. In particular, for his invaluable help and advice, I should like to thank Dr John Phillipson, Reader in Animal Ecology, University of Oxford. I should also like to thank: Dr S. McNeill, Imperial College, London, Dr J. Satchell, Institute of Terrestrial Ecology, Natural Environment Research Council, and Dr M. Swift, Queen Mary College, London.

R. H. Hedley   Director
British Museum (Natural History)
October 1978

Chapter 1

What goes on behind the scenes?

In this book you can take a look behind the scenes in an oak woodland and a rocky seashore. You can discover a different way of thinking about your natural surroundings.

The inhabitants

What kinds of plants live in an oak woodland . . .

and on a rocky seashore?

What kinds of animals live in an oak woodland . . .

and on a rocky seashore?

These are some of the more familiar
ones. How many more can you think
of?

*Did you think of all the plants shown
on page 24 and all the animals on
pages 32 and 36?*

The range of plants and animals you can see in this book only hints at the diversity of life in oak woodlands and on rocky seashores.

As well as the more familiar plants, animals and fungi, there are many that you have to search for. They may be hidden in the soil or sand, under bark or seaweed, or even inside other plants or animals.

hidden under bark

hidden under seaweed

In the sea, there are countless numbers of plants (phytoplankton) and animals (zooplankton) that are too small to be seen without a microscope.

And there are millions of microscopic bacteria, fungi and single-celled organisms living almost everywhere.

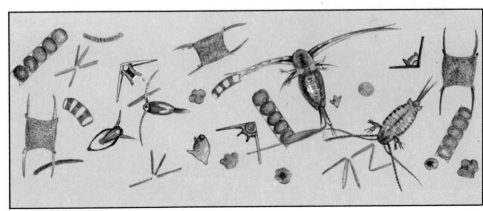

too tiny to be seen without a microscope

What do the inhabitants do?

They move, feed, sense changes in their surroundings, get rid of waste, respire, grow and reproduce. These are all **life processes.**

Sense changes in their surroundings – so they are aware of the activities of other animals and of changes in climatic conditions.

Get rid of waste that would be harmfu[l] if it accumula[tes] in their bodies.

Feed, and digest their food – to get the energy and raw materials they need to carry out their life processes.

Move – to search for food, escape from danger.

Respire – to release energy from their food by a complex series of chemical reactions.

Plants never look as if they are doing very much, but all the time they are feeding, respiring, getting rid of waste and sensing changes in their surroundings.

Plants do not move in the same sense that animals do, but parts of plants may move in response to changes in their surroundings. Flowers and leaves often close up at night.

Grow and reproduce

Plants and animals **grow** by building up new living material.

Seeds and eggs grow into adult plants and animals. A tiny acorn may grow into a tree 30 metres high; a caterpillar may double its size in two weeks. And plants and animals constantly **repair** damaged and worn out parts of themselves. A worm that is cut in half can grow a completely new half. When a branch breaks off a tree, the tree grows an area of callous tissue to seal off the wound and protect it from infection.

Plants and animals **produce new individuals** which replace them when they die.

11

Everything that 'works' needs a driving force – energy

Plants and animals work as they carry out their life processes. So they need a constant supply of energy.

Where do these working things get the energy they need?

Energy can take many forms . . .

heat energy

light energy

stored chemical energy

mechanical energy

. . . and can be converted from one form to another

When you strike a match, the stored chemical energy in the match-head is converted into heat energy and light energy.

Every time energy is converted from one form to another, some is converted into heat . . .

In an electric light bulb, electrical energy is converted into light energy – and heat energy is produced as well.

. . . but the energy is not used up – it just changes its form.

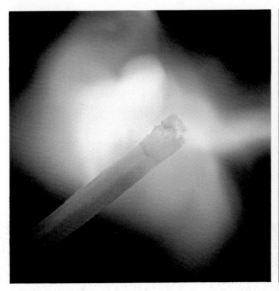 

Where do the inhabitants get their energy from?

The sun is the primary source of energy for all living things.

How do plants capture energy from the sun?
You can find out in Chapter 2.

And how do animals get the energy they need?
You can find out in Chapter 4.

The inhabitants need raw materials

In order to carry out their life processes, they need raw materials as well as energy.

The non-living surroundings are the primary source of raw materials for all living things.

Other aspects of the non-living surroundings, such as climatic conditions and the lie of the land, may also have an important influence on the lives of plants and animals.

How do plants take in raw materials from the non-living surroundings? *You can find out in Chapter 3.*

And how do animals get the raw materials they need? *You can find out in Chapter 4.*

The coloured dots represent raw materials.

If we choose an area and study . . .

all the **plants**

all the **animals**

and the **non-living surroundings . . .**

. . . we can find out how the plants and animals and other living things interact with each other and with their non-living surroundings to form a natural system – an **ecosystem.**

Chapter 2
Capturing the sun's energy

The sun is the primary source of energy for all living things. The sun's energy reaches the Earth after a journey of 150 million kilometres through space. As the energy passes through the Earth's atmosphere, as much as half of it may be absorbed or reflected back into space.

The energy that reaches the Earth's surface is mainly in the form of **light energy** and **heat energy.**

Living things capture some of this energy and use it to drive their life processes.

Heat energy cannot be captured by plants or animals, but it warms up all living things and their non-living surroundings.

Light energy can be captured, but only by green plants.

Green plants capture light energy and convert it into stored chemical energy which they use to drive their life processes.

We call green plants **autotrophs,** which means 'self-nourishing'

The energy-capturer – chlorophyll

Green plants are able to capture the sun's light energy because they contain an energy-capturing substance called **chlorophyll.**

Chlorophyll is green – this is why so many plants are green.

Although some plants, such as copper beech and many seaweeds, do not look green, they still contain chlorophyll. So they are able to capture the sun's light energy.

Their greenness is hidden by other pigments.

'Non-green' green plants

'Making things with light'
The light energy that chlorophyll captures is used to build simple raw materials into complex, energy-storing materials called carbohydrates. This process is called photosynthesis, which means 'making things with light'.

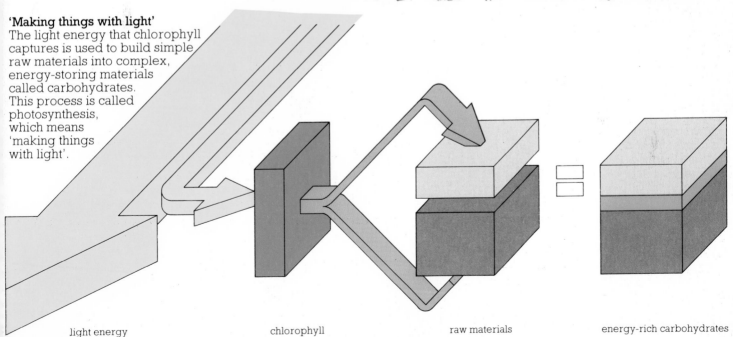

light energy   chlorophyll   raw materials   energy-rich carbohydrates

What raw materials are needed for photosynthesis?

The raw materials needed for photosynthesis are the raw materials that make up carbohydrates. As the name suggests, **carbo-hydr-ates** are made up of carbon, hydrogen and oxygen. (The ending '-ates' means that a chemical contains lots of oxygen.)

Plants get their carbon (C) and oxygen (O_2) from carbon dioxide (CO_2), and their hydrogen (H_2) from water (H_2O).

Most photosynthesis takes place in the leaves of green plants. How do the raw materials get to the leaves?

- The **carbon dioxide** enters the leaf through pores in the leaf surface called stomata.

- **Water** from the soil enters the plant through its roots. It is transported in special channels from the roots up the stem and into the veins of the leaves.

Plants that live in water are surrounded by dissolved raw materials. They take in water and carbon dioxide all over their surfaces.

Plants need a continual supply of carbon dioxide and water for photosynthesis, but chlorophyll can be used over and over again. Once chlorophyll has passed on its captured energy, it is ready to capture more.

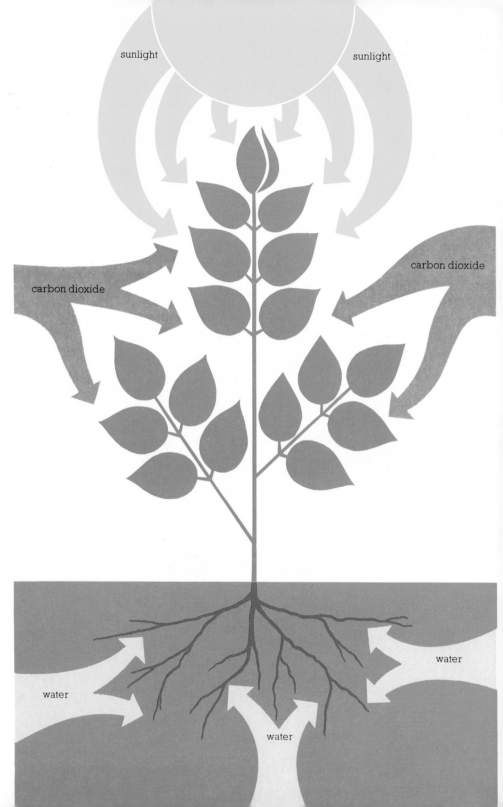

Where does photosynthesis take place?

Inside many of the cells in the leaves of green plants are lens-shaped structures called **chloroplasts.** Chloroplasts are very small, less than 0.004 of a millimetre across. They contain chlorophyll and are the site of photosynthesis.

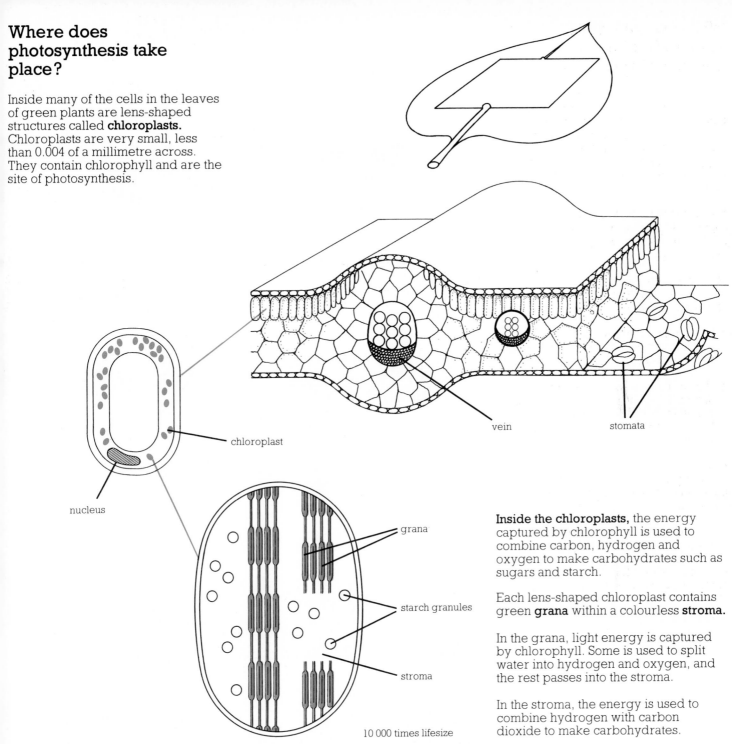

chloroplast

nucleus

vein

stomata

grana

starch granules

stroma

10 000 times lifesize

Inside the chloroplasts, the energy captured by chlorophyll is used to combine carbon, hydrogen and oxygen to make carbohydrates such as sugars and starch.

Each lens-shaped chloroplast contains green **grana** within a colourless **stroma.**

In the grana, light energy is captured by chlorophyll. Some is used to split water into hydrogen and oxygen, and the rest passes into the stroma.

In the stroma, the energy is used to combine hydrogen with carbon dioxide to make carbohydrates.

21

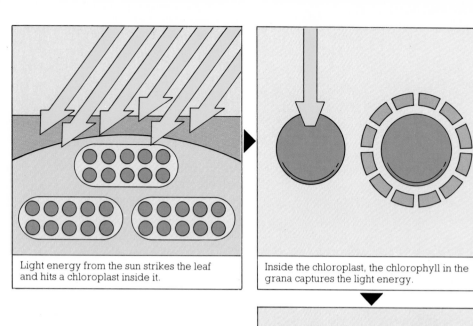

Light energy from the sun strikes the leaf and hits a chloroplast inside it.

Inside the chloroplast, the chlorophyll in the grana captures the light energy.

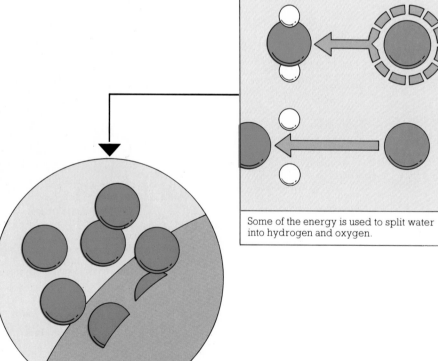

Some of the energy is used to split water into hydrogen and oxygen.

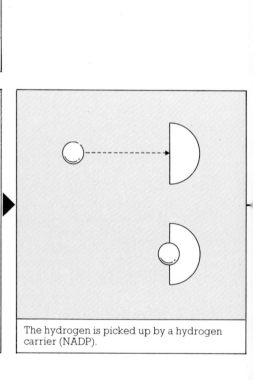

The hydrogen is picked up by a hydrogen carrier (NADP).

The oxygen is released into the air.

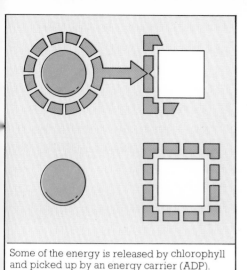

Some of the energy is released by chlorophyll and picked up by an energy carrier (ADP).

More about photosynthesis

These diagrams will help you understand what happens inside a chloroplast.

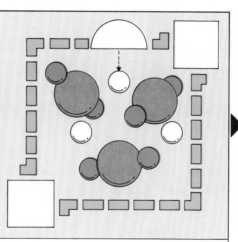

The carriers take hydrogen and energy into the **stroma**. Here some of the energy is used to combine hydrogen and carbon dioxide...

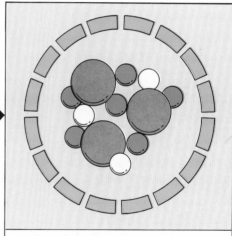

to make carbohydrates that are rich in stored chemical energy.

The energy-rich carbohydrates are carried round the plant to the cells where energy is needed to drive the plant's life processes.

Inside the cells, some of the carbohydrates are broken down to release their stored energy. This process of releasing stored energy is called **respiration**. Some carbohydrates are stored in the plant's cells for future use.

Carbon dioxide enters the leaf.

Photosynthesis is the process by which all green plants get the energy they need to drive their life processes. *You can find out the names of the plants on these pages on page 80*

Animals cannot capture the sun's light energy and use it to drive their life processes.

How do animals get the energy they need?
You can find out in Chapter 4

Chapter 3

Inside the growing plant

During photosynthesis, carbon, hydrogen and oxygen are built into energy-rich carbohydrates. But plants are made up of more than just carbohydrates. They need many raw materials to build up new plant material and carry out all their life processes.

The raw materials that plants need are in the non-living surroundings.

Plants are made up of many raw materials.

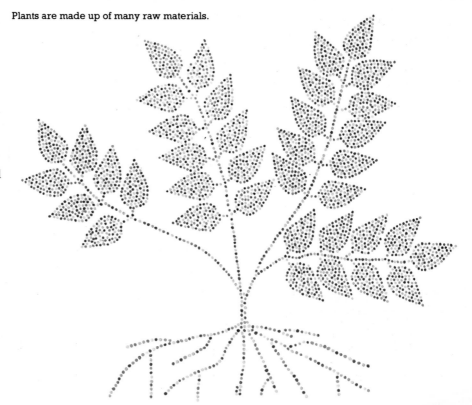

How do raw materials enter the plant?

Oxygen and carbon dioxide enter land plants through pores in their leaves and stems. All the other raw materials are taken in from the soil through their roots.

Plants that live in water are surrounded by dissolved raw materials, which they take in all over their surfaces.

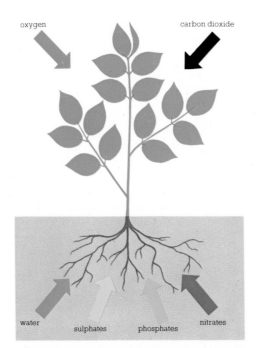

oxygen

carbon dioxide

water

sulphates

phosphates

nitrates

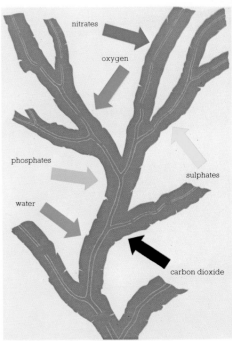

nitrates

oxygen

phosphates

sulphates

water

carbon dioxide

How do raw materials become part of the plant?

Every living plant cell uses energy to combine carbohydrates (from photosynthesis) and raw materials (from the non-living surroundings) to make new living material – to **grow**.

As a plant grows, it builds up many different materials which:

● carry out the plant's life processes

● support and protect the plant

● store energy that can be used to drive the plant's life processes.

All plant materials contain energy, but some materials are particularly rich in it. During respiration, these energy-storing materials can be easily broken down to release their stored energy. They are the only materials the plant uses to drive its life processes.

Materials that carry out life processes (such as the proteins in chlorophyll and enzymes)

These materials are found in every living plant cell. Many of them are made by combining carbohydrates with other raw materials, especially nitrates and phosphates.

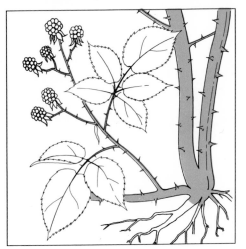

Materials for support and protection (such as the cellulose in cell walls and the lignin in wood)

In land plants, these materials make up the plant's 'skeleton'. They are made by combining many carbohydrates, and they are difficult to break down. Plants that live in water do not need a strong 'skeleton', because they are supported by the water.

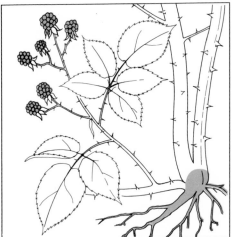

Materials for storing energy (such as starch, vegetable oils and fats)

In land plants, these materials may be stored in special structures such as bulbs and tubers. And most seeds are full of energy-storing materials. Plants living in the sea do not need to build up large reserves of energy-storing materials. This is because they do not have to survive such wide seasonal variations as land plants.

Growing . . .

As plants grow, new materials accumulate in their cells; the cells enlarge and eventually divide. So the number of cells in the plant increases.

In most plants, the cells remain attached to each other and the plant grows bigger.

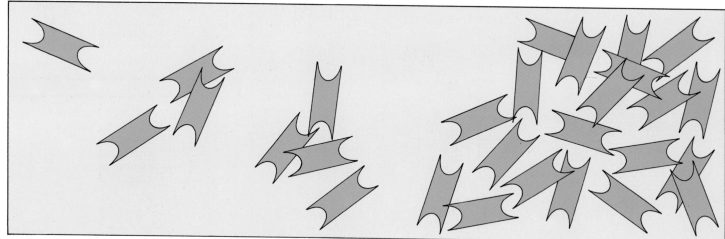

In single-celled plants, like these microscopic phytoplankton, the cells separate to form new individuals.

Climate, soil and plant growth

In all ecosystems, many aspects of the non-living surroundings affect the rate of plant growth. Three of the most important are temperature, rainfall and the supply of raw materials in the soil.

In the diagrams on this page you can see how these factors may interact.

But the real situation is much more complicated than this. The factors interact in a complex way, and different plants grow best in different conditions.

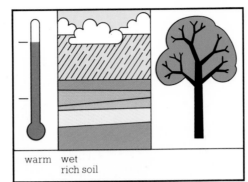

warm wet
rich soil

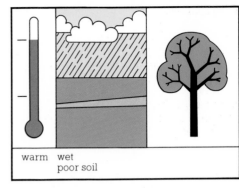

warm wet
poor soil

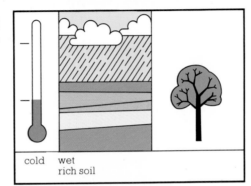

cold wet
rich soil

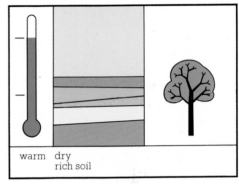

warm dry
rich soil

cold wet
poor soil

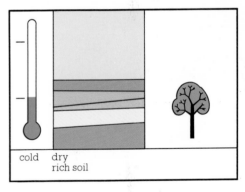

cold dry
rich soil

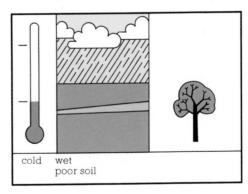

warm dry
poor soil

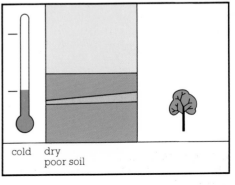

cold dry
poor soil

Chapter 4
Energy from food

Animals need **energy** and many **raw materials** in order to grow, reproduce and carry out all their other life processes.

Animals cannot capture and use the sun's light energy directly. So how do they get the energy they need?

Apart from oxygen and water, animals take in very few raw materials directly from their non-living surroundings. So how do they get the raw materials they need?

Animals get the energy and raw materials they need by eating plants . . . or other animals that have eaten plants.

Living things that get the energy and raw materials they need by feeding on other living things are called **consumers.**

We call consumers **heterotrophs**, which means 'getting nourishment from others'.

Consumers that feed mainly on plants are called herbivores

There is a wide range of plant material available to woodland herbivores – leaves, fruits, bark and plant juices such as sap and nectar.

There is less variety of plant material on a rocky shore – lichens, seaweeds and microscopic phytoplankton in the water.

Some woodland herbivores feed on many different plant materials. Others, such as this tiny **nut weevil** are much more selective.

Most seashore herbivores, such as this **limpet,** feed on young seaweeds. Others feed on phytoplankton.

Consumers that feed mainly on other animals are called carnivores

Carnivores are often called 'predators' and the animals they feed on are called their 'prey'.

Some carnivores, such as this **tawny owl,** prey mainly on herbivores.

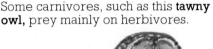

Some carnivores, such as this **rock goby,** prey mainly on other carnivores.

Consumers that feed on plant and animal material are called omnivores

Some omnivores strain their food from sea water and can select particles only by size, so they take in both plant and animal material.

Some omnivores change their diet as different kinds of food become available during the year.

Look at the range of consumers on pages 32 and 36.

Can you pick out the **herbivores,** the **carnivores,** and the **omnivores?**

Acorn barnacles strain particles of plant and animal material from sea water.

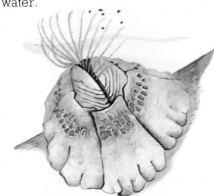

Badgers feed on small animals such as voles, frogs, snails and worms. They also eat grass, seeds and soft fruits.

Woodland consumers

You can find out the names of these animals and whether they are herbivores, carnivores or omnivores on page 78.

Before a consumer can use the energy and raw materials in its food, it must find its food, take it in, break it down, absorb it and transport it to the parts of its body where it is needed.

Finding food

Some carnivores, such as **foxes** and **herring gulls,** spend a great deal of time and energy moving around in search of food.

Other carnivores, such as **spiders** and **sea anemones,** lie in wait for their prey, so they use less energy finding food.

Most herbivores, such as this **mottled umber moth caterpillar** or **topshells,** live on or very near to their food, so they spend very little time and energy searching for it.

Taking in food

Consumers have mouthparts that are suited to taking in the particular kind of food they eat.

Peacock fanworms have long tentacles which extend into the water to catch particles of plant and animal material.

Butterflies and **moths** have a long, coiled proboscis which they extend to reach inside flowers and suck out the nectar.

Voles have sharp front teeth for nibbling at tough plant material.

Sparrowhawks use their sharp, curved beaks to tear at their prey.

Breaking down food

All consumers must break down their food into a form that can be absorbed into their bodies. This process is called **digestion** and in many consumers it takes place inside a gut.

During digestion, food is broken down into smaller and smaller particles until it will dissolve in water. In this dissolved form, the digested food can be absorbed into the consumer's body and transported to all the cells where it is needed.

In the cells of the consumer's body, some food materials are broken down to release stored energy, which is used to drive life processes. Other food materials are recombined with each other to form new living material as the animal grows. So, all over the consumer's body, there are materials containing stored chemical energy.

Digestion in herbivores
Most land plants contain a lot of woody supporting material (particularly cellulose), but most herbivores cannot produce the appropriate enzymes to break down this material. They rely on bacteria and other minute organisms living in their gut to digest the cellulose for them.

Digestion outside the body
Some consumers, such as bacteria and fungi, have no gut. They pass out enzymes on to their food and absorb the digested food directly into their bodies.

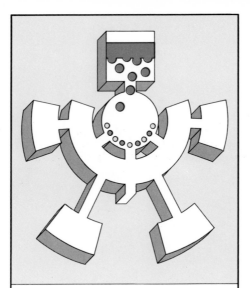

Mechanical breakdown, for example by the grinding back teeth of rabbits and voles, or the crushing jaws of the ballan wrasse.

Chemical breakdown with the help of digestive enzymes. There are many different digestive enzymes, each specialized to break down a particular kind of food. The enzymes themselves are not broken down during digestion, so they can be used more than once.

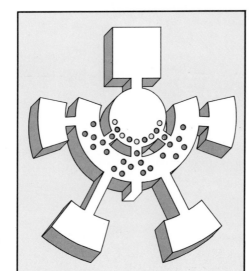

Absorbing digested food through the wall of the gut.

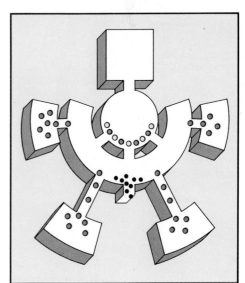

Transporting digested food all over the body. **Getting rid of waste**
Food that is not digested, perhaps because the consumer cannot make the correct enzymes, passes out of the body as solid waste. What happens to this waste material? *You can find out on page 38.*

Seashore consumers

You can find out the names of these animals and whether they are herbivores, carnivores or omnivores on page 79.

Waste not, want not . . .

Undigested food and the dead bodies of plants and animals still contain energy and raw materials. Other consumers, called **decomposers**, can feed on this dead and waste material. As they feed, they break it down into smaller and smaller fragments.

Many decomposers can digest only a small amount of the material they take in, so they have to get rid of a lot of waste. **Bacteria** and **fungi** are special types of decomposer that can carry on where others leave off. They eventually convert all dead and waste plant and animal material into simple chemicals, which may be used as raw materials by plants.

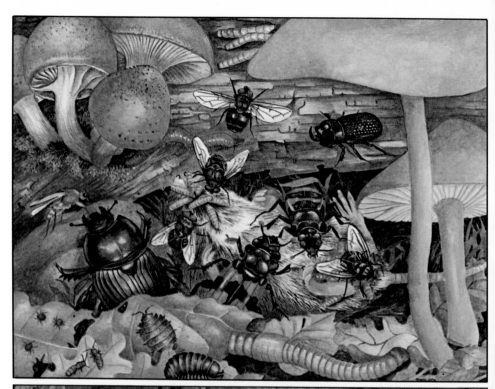

. . . the importance of decomposers

At least 80 per cent of plant material in woodlands is broken down by decomposers.

The entire process of decomposition may take many years, but if there were no decomposers we would be surrounded by piles of natural litter!

You can find out the names of the decomposers in these pictures on page 81.

The action of waves and tides begins the breakdown of dead plants and animals on the seashore.
Decomposers feed on the debris and break it down into smaller and smaller fragments. When the fragments are very small, decomposers that sieve debris from the sand and water may feed on them.

Chapter 5

Food chains

When a herbivore eats a plant . . . and then a carnivore eats the herbivore . . . the sequence of events is called a **food chain.**

We can think of each plant or animal as a link in the chain, so this is a three-link food chain.

In the world around us there are millions of these simple three-link food chains involving a plant, a herbivore and a carnivore.

Four-link food chains are also very common. In a four-link food chain, the first carnivore is eaten by a second carnivore, which is usually larger and more aggressive.

Food chains in an oak woodland and on the seashore

In this picture (based on a coastal area of Cornwall) you can see the plants and animals which make up some common three and four-link food chains.

How many food chains can you work out?
Some of them are shown on the next page.

40

Did you find all these woodland food chains?

Foxglove flowers → Bumble bee → Jay

Blackberries → Bank vole → Stoat

Oak leaves → Winter moth caterpillar → Robin

Lichen → Great grey slug → Common toad → Grass snake

Oak leaf sap → Oak leaf aphid → Great tit → Sparrowhawk

Chanterelle → Fungus beetle → Common frog → Badger

Did you find all these seashore food chains?

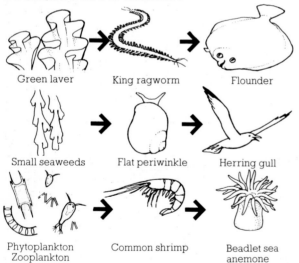

Green laver → King ragworm → Flounder

Small seaweeds → Flat periwinkle → Herring gull

Phytoplankton Zooplankton → Common shrimp → Beadlet sea anemone

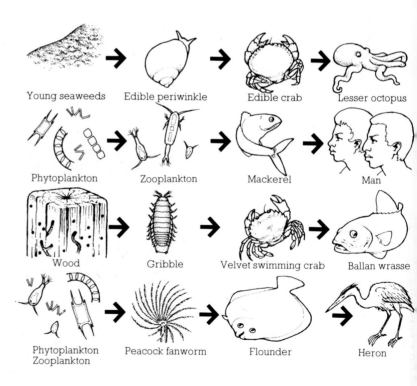

Young seaweeds → Edible periwinkle → Edible crab → Lesser octopus

Phytoplankton → Zooplankton → Mackerel → Man

Wood → Gribble → Velvet swimming crab → Ballan wrasse

Phytoplankton Zooplankton → Peacock fanworm → Flounder → Heron

The first consumer in a food chain is not always a herbivore. On the seashore many consumers are omnivores because they feed by filtering both plants (phytoplankton) and animals (zooplankton) from the sea water.

Waste not, want not . . . decomposer food chains

Animal waste materials and the dead bodies of plants and animals are food for **decomposers.** They form the first links in **decomposer food chains.**

Can you pick out the two decomposer food chains shown in these pictures?

You can find out what they are on page 83.

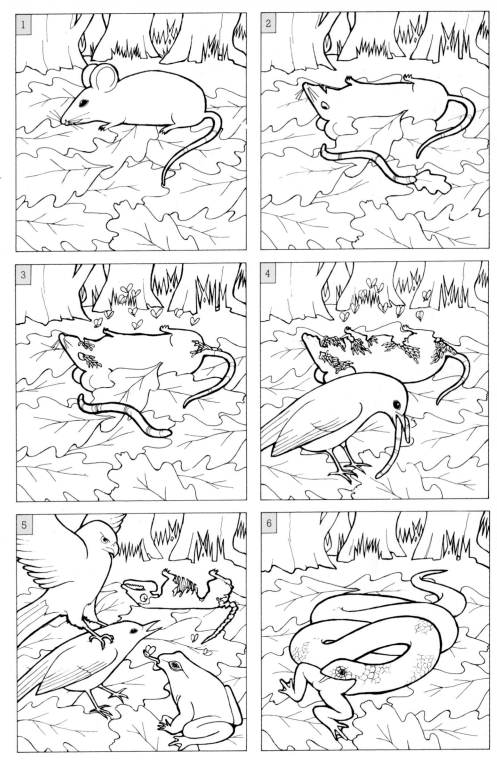

Energy transfer along a food chain

Each time something is eaten, energy-containing materials are transferred from the food to the feeder. This is how energy is transferred along a food chain.

first transfer – from the leaf to the caterpillar

second transfer – from the caterpillar to the shrew

third transfer – from the shrew to the badger

As energy-containing materials are transferred along a food chain, the amount that is transferred gets smaller and smaller because some is **wasted** and some is **used** by each consumer in the chain.

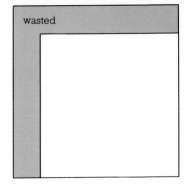

energy-containing
materials in the leaf
eaten by the
caterpillar

Of the energy-containing materials transferred from the leaf to the caterpillar . . .

Some are **wasted** because the caterpillar can't digest all the leaf material it eats. The undigested material passes out of the caterpillar's body as solid waste.

This waste material forms the first link in decomposer food chains.

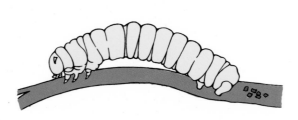

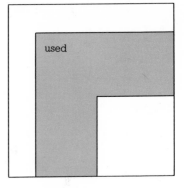

wasted

Some are **used** to drive the caterpillar's life processes. As energy drives life processes it is converted into heat energy, which is released into the non-living surroundings and lost from the food chain.

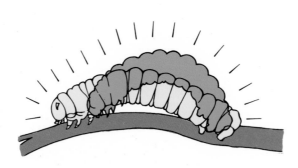

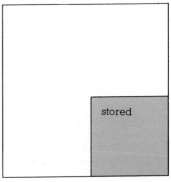

used

Energy-containing materials that are neither wasted nor used are recombined and **stored** as new living material in the caterpillar's body.

. . . this stored energy is all that is available to the next consumer in this food chain.

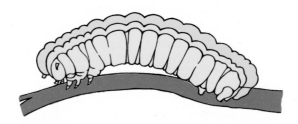

stored

Recycling raw materials

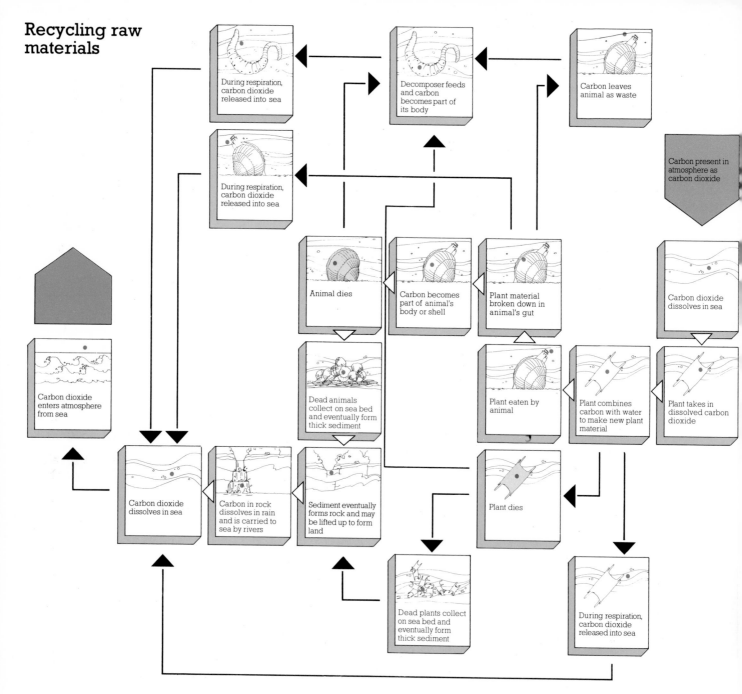

During respiration, carbon dioxide released into sea

Decomposer feeds and carbon becomes part of its body

Carbon leaves animal as waste

Carbon present in atmosphere as carbon dioxide

During respiration, carbon dioxide released into sea

Animal dies

Carbon becomes part of animal's body or shell

Plant material broken down in animal's gut

Carbon dioxide dissolves in sea

Carbon dioxide enters atmosphere from sea

Dead animals collect on sea bed and eventually form thick sediment

Plant eaten by animal

Plant combines carbon with water to make new plant material

Plant takes in dissolved carbon dioxide

Carbon dioxide dissolves in sea

Carbon in rock dissolves in rain and is carried to sea by rivers

Sediment eventually forms rock and may be lifted up to form land

Plant dies

Dead plants collect on sea bed and eventually form thick sediment

During respiration, carbon dioxide released into sea

The materials transferred along food chains are built up by green plants from energy which they capture and raw materials which they take in from their surroundings.

The sun provides plants with a continuous supply of energy, but the supply of raw materials on the Earth is limited.

Why doesn't the supply of raw materials run out?

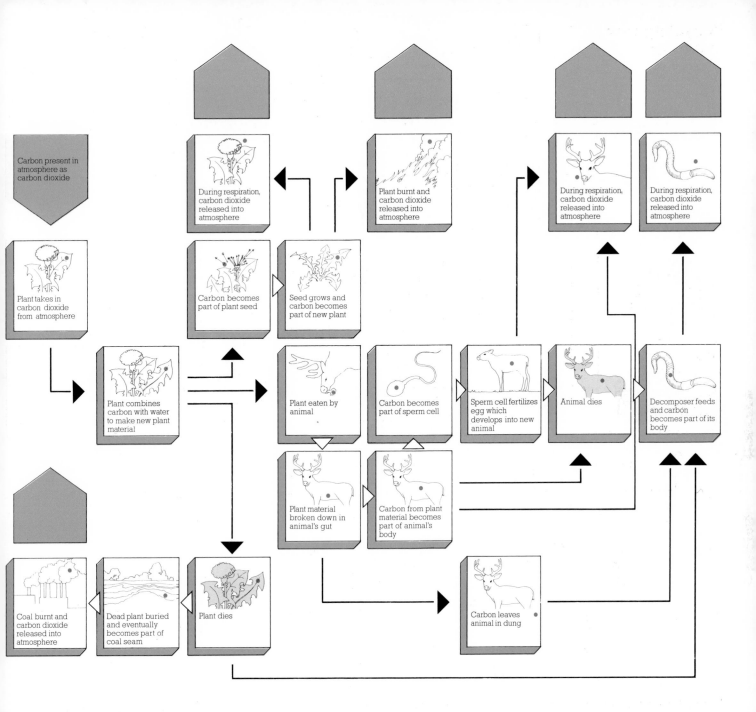

Carbon present in atmosphere as carbon dioxide

Plant takes in carbon dioxide from atmosphere

Plant combines carbon with water to make new plant material

During respiration, carbon dioxide released into atmosphere

Carbon becomes part of plant seed

Seed grows and carbon becomes part of new plant

Plant burnt and carbon dioxide released into atmosphere

During respiration, carbon dioxide released into atmosphere

During respiration, carbon dioxide released into atmosphere

Plant eaten by animal

Carbon becomes part of sperm cell

Sperm cell fertilizes egg which develops into new animal

Animal dies

Decomposer feeds and carbon becomes part of its body

Plant material broken down in animal's gut

Carbon from plant material becomes part of animal's body

Coal burnt and carbon dioxide released into atmosphere

Dead plant buried and eventually becomes part of coal seam

Plant dies

Carbon leaves animal in dung

The supply doesn't run out because the atoms which make up the raw materials are used over and over again – **they are recycled.**

On these pages you can follow some of the pathways that a carbon atom may take as it is recycled. Try to find the longest pathway and the shortest one.

It may be weeks, months or even millions of years before the carbon atom returns to the atmosphere.

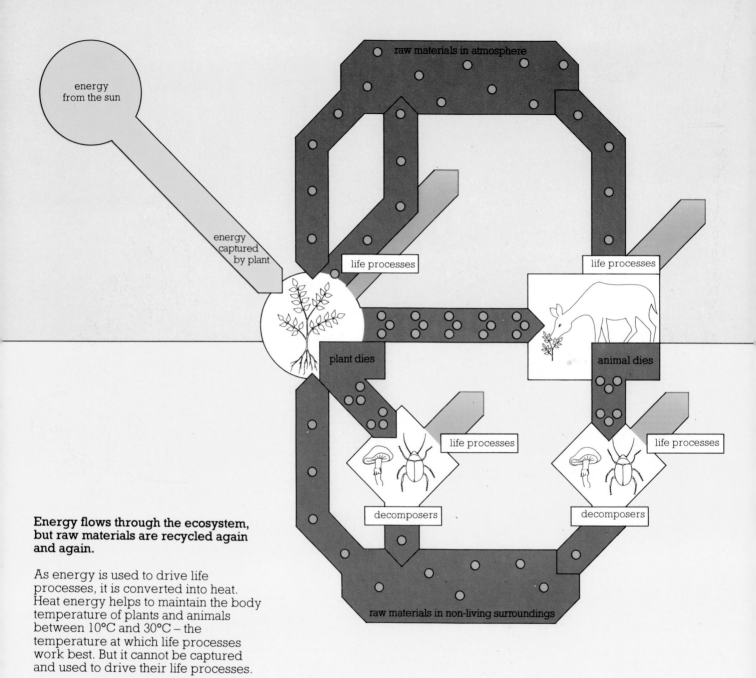

energy from the sun

energy captured by plant

raw materials in atmosphere

life processes

life processes

plant dies

animal dies

life processes

life processes

decomposers

decomposers

raw materials in non-living surroundings

Energy flows through the ecosystem, but raw materials are recycled again and again.

As energy is used to drive life processes, it is converted into heat. Heat energy helps to maintain the body temperature of plants and animals between 10°C and 30°C – the temperature at which life processes work best. But it cannot be captured and used to drive their life processes.

So energy cannot be recycled. Plants and animals depend on a continuous supply of energy from the sun to drive their life processes.

Chapter 6

Food webs

Each consumer in a food chain 'wastes' or 'uses' nearly all of the energy-containing materials it takes in. Only a few per cent are taken in by the next consumer in the chain.

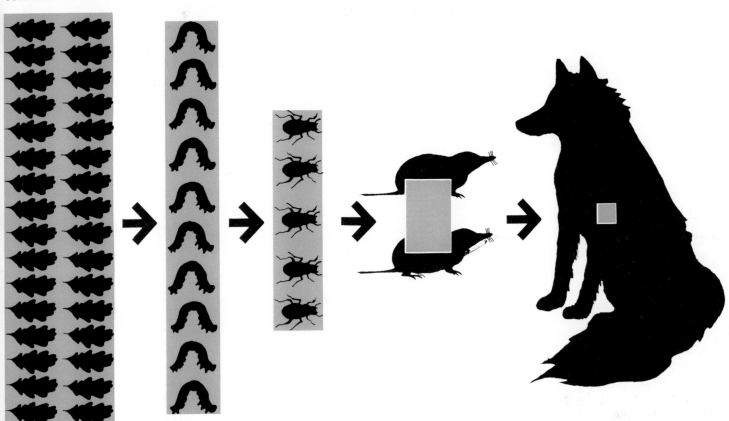

So, at the beginning of a food chain, there are enough energy-containing materials to support a **large number of herbivores**. But after three or four transfers there are only enough to support a very **small number of carnivores**.

Could there be a sixth link in this chain?

It seems unlikely because:
- there are very few energy-containing materials left after three transfers
- the next carnivore in the chain would have to be large enough to kill the fox, and large animals need a lot of food.

This may explain why most food chains have fewer than six links.

Longer food chains

Most food chains have only three or four links, but there are some longer food chains.

How is this possible?

Longer food chains can exist because most carnivores do not rely on the animal immediately before them in the chain to provide all the food they need.

The carnivores at the end of longer food chains have varied diets.

They may vary their diet by feeding at different positions in the same chain . . .

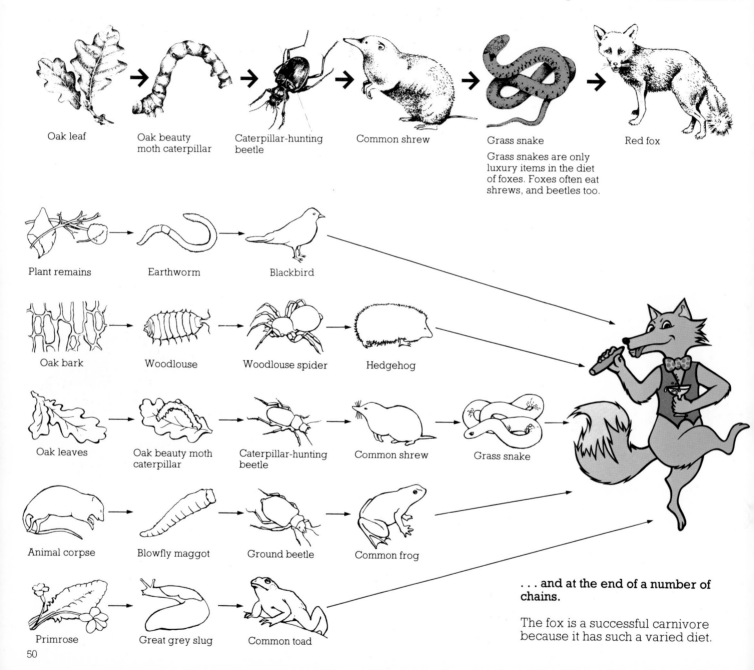

Oak leaf → Oak beauty moth caterpillar → Caterpillar-hunting beetle → Common shrew → Grass snake → Red fox

Grass snakes are only luxury items in the diet of foxes. Foxes often eat shrews, and beetles too.

Plant remains → Earthworm → Blackbird

Oak bark → Woodlouse → Woodlouse spider → Hedgehog

Oak leaves → Oak beauty moth caterpillar → Caterpillar-hunting beetle → Common shrew → Grass snake

Animal corpse → Blowfly maggot → Ground beetle → Common frog

Primrose → Great grey slug → Common toad

. . . and at the end of a number of chains.

The fox is a successful carnivore because it has such a varied diet.

50

Interconnecting food chains

Animals earlier in food chains may also vary their diets – they may feed in more than one chain.

This allows them to exploit alternative food sources when their main food becomes scarce, and to take advantage of gluts of food in other food chains.

So we find that there are many connections between food chains.

In the diagram below some of the connections are drawn in for you. Try to draw the others yourself. The list below tells you what each animal feeds on.

Grass snakes eat frogs and small mammals, such as shrews.

Hedgehogs feed mainly on spiders and insects but they sometimes eat grass snakes too.

Frogs eat a variety of small animals, such as insects and worms.

Caterpillar-hunting beetles eat maggots as well as caterpillars.

Oak beauty moth caterpillars feed on oak leaves.

Earthworms feed on the remains of dead plants.

Shrews eat spiders and many small insects.

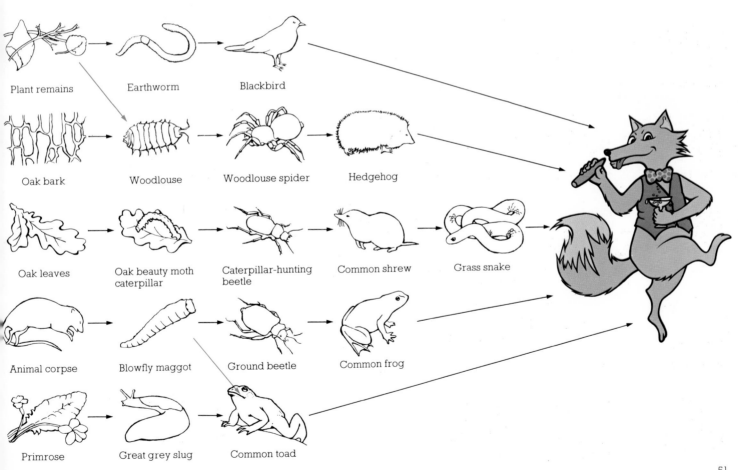

Plant remains   Earthworm   Blackbird

Oak bark   Woodlouse   Woodlouse spider   Hedgehog

Oak leaves   Oak beauty moth caterpillar   Caterpillar-hunting beetle   Common shrew   Grass snake

Animal corpse   Blowfly maggot   Ground beetle   Common frog

Primrose   Great grey slug   Common toad

. . . so how could this happen?

·STOP PRESS·

SCIENTIST'S EXPLANATION OF BAFFLING BANKRUPTCY

An Agricultural spokesman revealed today that the greatly reduced number of rabbits due to myxomatosis had led to starving foxes attacking and eating young lambs in many sheep-farming areas of the country. The resulting scarcity of lamb throughout the country had led to a dramatic fall in the sale of mint sauce, and subsequently to the demise of the already ailing empire of Mr Spear Mentha.

MYXOMATOSIS BANKRUPTS MINT SAUCE TYCOON

Woodland food web

There are so many connections between food chains that we can think of every organism as part of a complex **food web,** rather than as a link in a straight chain.

A change in the number, or feeding habits, of organisms at one link in a food web may affect many other plants and animals in the web.

So man's efforts to wipe out a plant or animal harmful to his economy must be planned very carefully to avoid upsetting the balance of a food web.

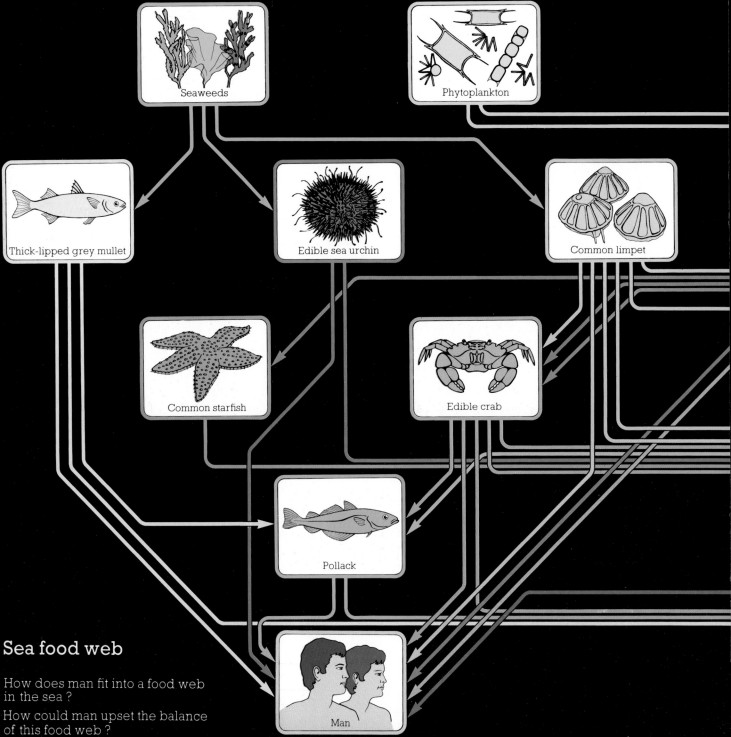

Sea food web

How does man fit into a food web in the sea ?

How could man upset the balance of this food web ?

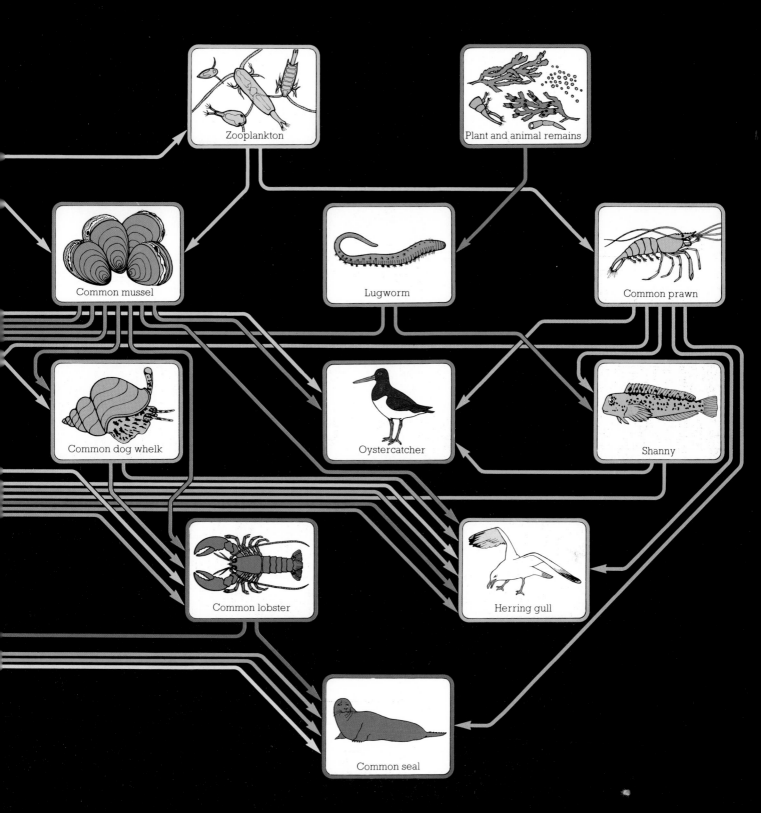

Chapter 7
Building pyramids

In the last two chapters you have seen how energy is transferred along individual food chains. But how can we understand the way energy flows through a complex food web or a whole ecosystem?

Grouping into feeding levels

We can start by grouping plants and animals into feeding levels which we call **trophic levels** (*trophe* is a Greek word which means nourishment).

How do we know which trophic level a plant or animal belongs to?

Green plants are at trophic level 1 because their energy has been transferred once:
● from the sun to plants.

Herbivores are at trophic level 2 because their energy has been transferred twice:
● from the sun to plants
● from plants to herbivores.

Carnivores that eat herbivores are at trophic level 3 because their energy has been transferred three times:
● from the sun to plants
● from plants to herbivores
● from herbivores to carnivores.

These carnivores that eat other carnivores are at trophic level 4 because their energy has been transferred four times:
● from the sun to plants
● from plants to herbivores
● from herbivores to carnivores
● from carnivores to second carnivores.

To find the trophic level, we count the number of energy transfers.

Did you notice . . .

. . . that robins may feed at more than one trophic level?

. . . and that tadpoles feed at a different trophic level from adult frogs?

Why is this?

Many animals vary their diet with the seasons. As one food source becomes scarce, they change to another.

In spring and summer, robins eat caterpillars and other insects. They are getting their energy by eating herbivores, so they are at trophic level 3.

In autumn and winter, robins eat seeds and berries. They are getting their energy by eating plants, so they are at trophic level 2.

Some animals change their diet as they grow up. The adults may have different mouthparts from the young, so they can feed on different kinds of food.

A change in diet can mean a change in trophic level.

How many animals can you think of that feed at more than one trophic level?

. . . but can you think of any plants?

Tadpoles eat water-weeds. They are getting their energy by eating plants, so they are at trophic level 2.

Frogs eat slugs, snails and insects. They are getting their energy by eating herbivores, so they are at trophic level 3.

In the woodland food web on page 53 how many animals are feeding at trophic level 2 ?
Which trophic level is the tawny owl at ?

Can you find an animal which is feeding at more than one trophic level ?
Turn to page 83 to see if you are right.

Measuring stored energy

Once we have grouped the plants and animals in a chosen area into trophic levels . . . the next step is to work out how much energy is stored by each trophic level over a whole year.

It would take too long to measure the energy stored in every plant and animal. (And we could destroy much of the ecosystem in the process.) So we usually choose **sample areas** and study these in detail. We use the results from these sample areas to estimate the amount of energy stored by each trophic level in the whole area.

Measuring energy directly is difficult and time consuming, but we can get a rough estimate of the amount of energy stored by counting or weighing the living things at each trophic level.

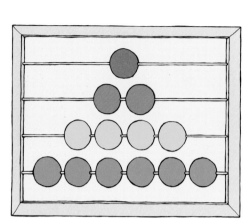

● **counting** the living things at each trophic level

● **weighing** the living things at each trophic level.

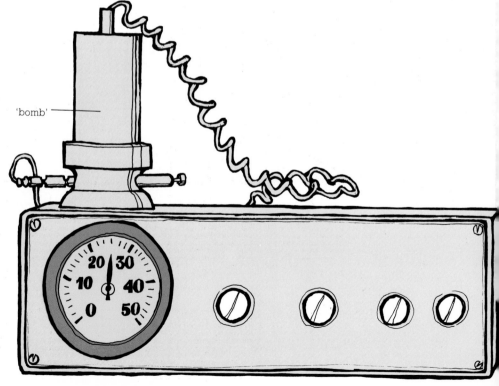

'bomb'

The only direct way of measuring the amount of energy stored in plants and animals is to use a **bomb calorimeter.** This is a complicated piece of equipment, but in principle this is how it works . . .

Inside the 'bomb' of the calorimeter, the stored energy in a sample of plant or animal material is converted to heat energy. The heat energy is measured as an increase in temperature and this is converted to an energy measurement. (Energy is measured in joules or calories.)

... throughout the year

Whichever way we measure stored energy, we need to take measurements many times throughout the year. We need to do this because **the amount of energy stored by plants and animals varies throughout the year.**

January–February

March–April

May–June

July–August

September–October

November–December

Light, temperature and rainfall affect the amount of energy stored by plants.

Woodmice eat mainly seeds, buds and insects. They store most energy (grow and reproduce) between April and September. At this time of year there is a good supply of plant food and they use very little energy keeping warm.

Tawny owls eat woodmice and other small animals. They lay their eggs in late March and the young hatch at the end of April. So the young owls are growing when there is a good supply of small animals for food.

Building a pyramid of energy

If we study an ecosystem for a whole year and estimate how much energy is stored by . . .

the plants at trophic level 1

the herbivores at trophic level 2

the carnivores at trophic level 3

the carnivores at trophic level 4

The result is what we call a pyramid of energy

This is the shape we would expect, because we know that each trophic level must provide enough energy to support the one above it.

Puzzling pyramids

If we estimate the amount of energy stored by **counting** all the living things at each trophic level . . .

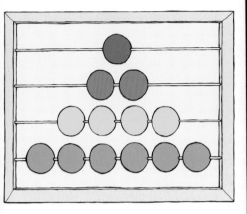

. . . the result is what we call a **pyramid of numbers.**

But sometimes counting gives us a puzzling result.

Look at the pyramid on the right. It looks as if there are not enough plants at trophic level 1 to provide the energy needed to support trophic level 2.

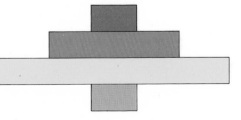

Although there is only one plant at trophic level 1, it is large enough to provide the energy needed to support the many small animals at trophic level 2.

Pyramids of numbers may be misleading, because they cannot take into account the size of things.

If we estimate the amount of energy stored by **weighing** all the living things at each trophic level

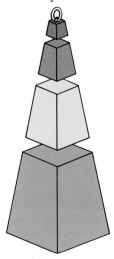

. . . the result is what we call a **pyramid of biomass.**

When we build pyramids, it is important to take measurements over a whole year.

The pyramid above the pictures is misleading because measurements were taken at only one time of year.

● Trophic level 1 is phytoplankton – they reproduce many times in a year.

● Trophic level 2 is barnacles – they reproduce only once or twice in a year.

Over a whole year, phytoplankton stores more energy than barnacles. So, if the measurements were taken over a year, we would see that the phytoplankton at trophic level 1 does provide enough energy to support the barnacles at trophic level 2.

All pyramids may be misleading if the measurements are taken at only one time of year, because they do not take into account the variation in amount of energy stored throughout the year.

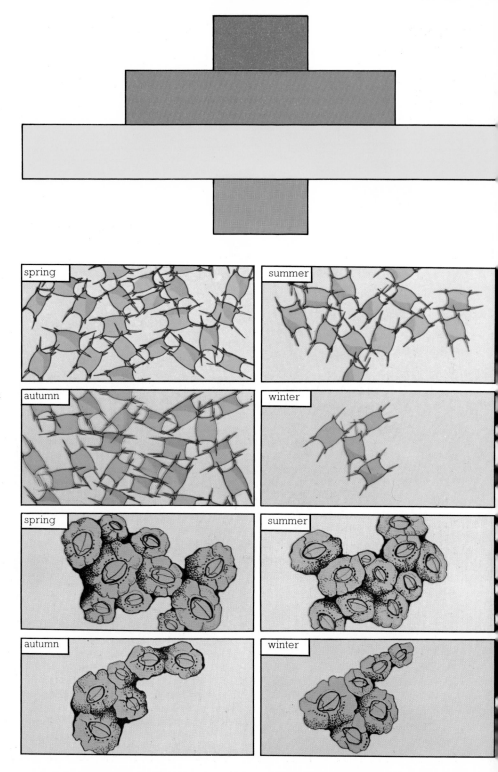

spring

summer

autumn

winter

spring

summer

autumn

winter

What about decomposers?

We can get some idea of the distribution of energy in an ecosystem by building pyramids. But to get a complete picture we would have to study the decomposers too.

It is difficult to get a good estimate of the amount of energy stored by decomposers over a year. Some of them are difficult to find and most of the true decomposers are too tiny to be seen without a microscope. We would also have to take very frequent measurements because bacteria reproduce many times in a day.

There is not enough information available for us to get an accurate picture of the amount of energy stored by decomposers over a year. But we do know that at least 80 per cent of the plant material in a woodland is broken down by decomposers.

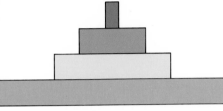

We must never underestimate the importance of decomposers in an ecosystem.

Chapter 8

Using pyramids

Why a pyramid? Of the energy that enters a trophic level . . .

● Some is **wasted** by the living things at that level. This energy is available to decomposers, but not to the animals at the next trophic level in the pyramid.

● Some energy is **used** by the living things to drive their life processes. As energy is used it is converted into heat (and lost), so it is not available to other living things.

● The small amount of energy that remains is **stored** by the plants or animals as they grow. **Only this stored energy is available to the animals at the next trophic level in the pyramid.**

Because energy is wasted and used by all living things in the pyramid, there is less energy at each successive trophic level. This is why we get a pyramid shape if we measure the amount of energy stored at each trophic level over a whole year.

The shape of a pyramid

The **shape** of a pyramid depends on **how much** energy is transferred from one trophic level to the next during a year.

This depends on how efficient the plants or animals at each trophic level are at:

● capturing energy from the trophic level below

● converting this energy into energy stored in their bodies.

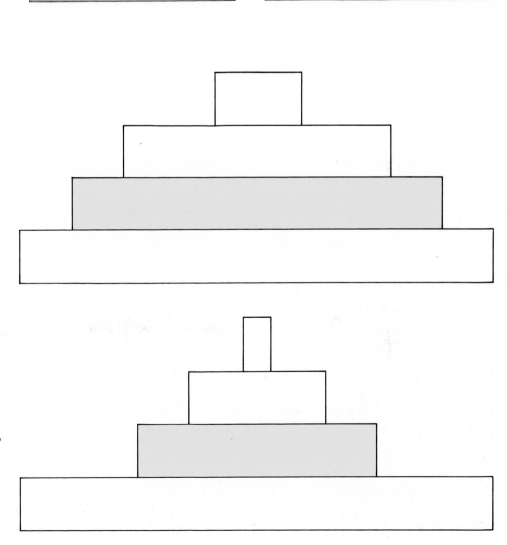

Which of these two pyramids has the more efficient herbivores?

You can find out on page 83.

By looking at the shape of the pyramid, we can compare the efficiency of the animals at different trophic levels.

By comparing the shapes of different pyramids, we can compare the distribution of energy in different ecosystems.

Comparing trophic levels

If you look at this woodland pyramid, you can see that the carnivores at trophic level 3 are more efficient than the herbivores at the trophic level below. Why do you think the carnivores are more efficient?

The pictures on these pages may give you some clues.

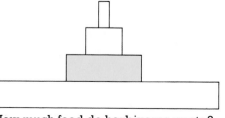

How much food do herbivores waste?

Some plant material is difficult to chew and swallow. Some contains a lot of cellulose and lignin, which are difficult to digest. So herbivores produce a lot of waste and they have to take in a lot of food to get the energy they need.

Animal material is easier to digest and more nutritious than plant material. So carnivores take in more of the energy in their food, and produce less waste than herbivores. Carnivores are therefore said to be more efficient.

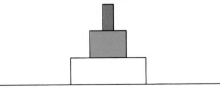

How much food do carnivores waste?

But remember, the waste materials produced by herbivores and carnivores are not really wasted – they are food for decomposers.

Comparing pyramids

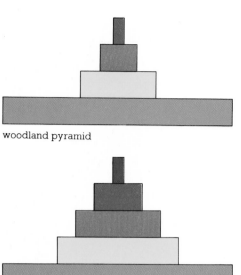

woodland pyramid

seashore pyramid

Let's look at the herbivores . . .

Seashore herbivores seem to be more efficient than woodland herbivores.

Why do you think this is?

Could one of the reasons be a difference in the plants they eat?

Plants living in the sea do not need a strong 'skeleton' to support them, so they contain less woody material than woodland plants.

Woody material is difficult to digest, so woodland herbivores produce more waste than seashore herbivores.

This may partly explain the difference in efficiency between seashore and woodland herbivores.

Let's look at the carnivores . . .

Seashore carnivores seem to be more efficient than woodland carnivores.

Why do you think this is?

Could one of the reasons be a difference in the amount of energy they use in capturing their food?

Some seashore carnivores are fairly inactive and slow moving. And some, such as the sea anemone, just wait for their food to come to them.

In contrast to this, many woodland carnivores use a lot of energy in moving around searching for their prey. Using lots of energy means that less is available to be stored. This may be one reason why woodland carnivores are less efficient than seashore carnivores.

Seashore animals tend to be more efficient than woodland animals. So, at each successive trophic level, there is more energy left in a seashore pyramid than in a woodland one.

The extra energy means that seashore food chains tend to be longer than woodland food chains, and seashore pyramids have more trophic levels than woodland ones.

So far, we have only compared the efficiency of herbivores and carnivores in the seashore and woodland ecosystems. But to get a complete picture of the distribution of energy in the two ecosystems, we would have to look at the decomposers too.

Chapter 9

Understanding ecosystems

We can study an ecosystem in terms of . . .

the plants

the animals

their non-living surroundings . . .

and begin to understand how all the different parts are related.

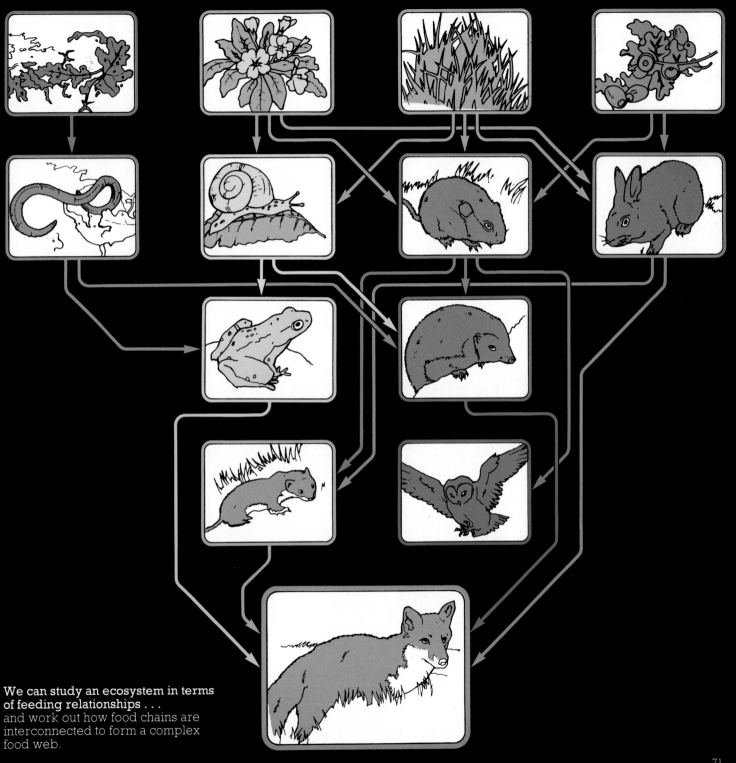

We can study an ecosystem in terms of feeding relationships . . .
and work out how food chains are interconnected to form a complex food web.

71

We can study an ecosystem in terms of trophic levels and pyramids.

If we measure the amount of energy stored at each trophic level over a year, we can build up a pyramid of energy.

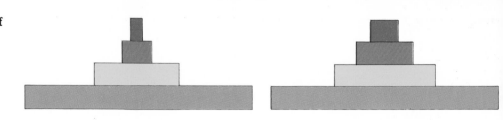

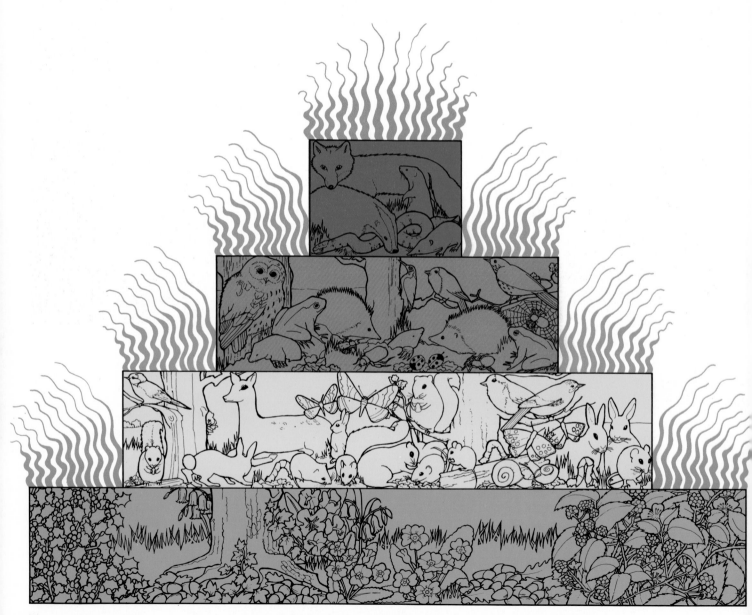

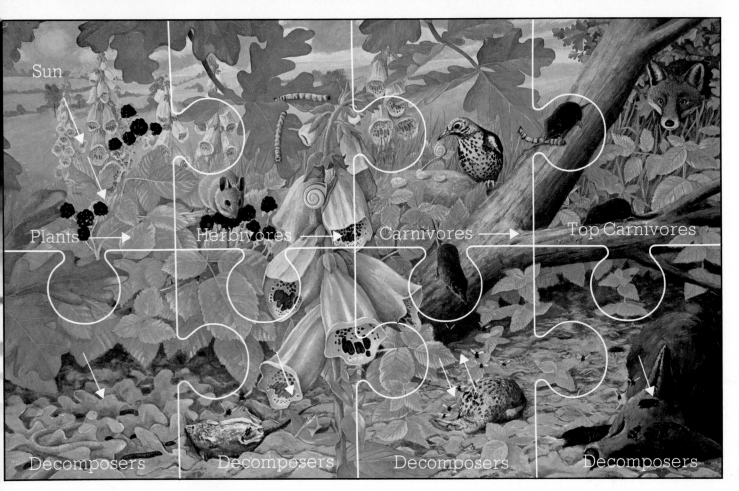

Sun

Plants → Herbivores → Carnivores → Top Carnivores

Decomposers Decomposers Decomposers Decomposers

We can study an ecosystem in terms of energy flow.

We can use food chains, trophic levels and energy measurements to build up a picture of how energy flows through the ecosystem.

What would happen . . .

if the sheep were fenced out of this oak woodland?

Small plants and ash and birch seedlings would start to appear.

Ash seedlings grow faster than oak seedlings. As the ash trees grow, ash leaf litter would begin to build up on the woodland floor.

Ash leaf litter is better food for earthworms than oak leaf litter so the number of earthworms would increase.

Earthworms make up more than half the diet of badgers. So how might these changes affect the population of badgers in the area?

You can find out on page 83.

Once you understand more about how ecosystems work you may begin to think differently about your natural surroundings.

74

Explore …
discover …
understand

This book can give you only a glimpse of **Nature at work.** You can discover much more by exploring for yourself.

Some Useful Books . . .

. . . to help you identify the living things you find

Collins Guides and **Collins Field Guides** are classic series of well-illustrated and comprehensive guides. They include guides to animal tracks and signs, birds, fishes, insects, life on the seashore, mammals, mushrooms and toadstools, trees, and many others.

Collins Handguides are much slimmer, less detailed, guides dealing with birds, butterflies and moths, fishes, trees, wild animals, and wild flowers.

Hamlyn Guides are another reliable series of guides covering the wildlife of Britain and Europe, including birds, freshwater fishes, trees, and organisms of the seashore and shallow seas.

Usborne Spotters Guides are an excellent series of small guides, including birds, birds of prey, butterflies, fishes, flowers, insects, mammals, mushrooms and other fungi, town and city wildlife, and trees. In *The Spotters Guide to Wildlife* and *The Spotters Handbook*, several guides are collected together.

The Tree Key by Herbert E. Edlin, Frederick Warne, 1978. A slightly different kind of identification guide. *The Wild Flower Key* and *The Bird Watchers Key* are also available.

Wild Flowers of Britain by Roger Phillips, Pan Books, 1977. A guide based on colour photographs. arranged in calendar order. Other books by the same author deal with fungi, trees, and grasses, ferns, mosses and lichens.

Wildflowers of Britain, Readers Digest Nature Lovers Library, 1981. In this guide, detailed illustrations are accompanied by small photographs of the flowers in their natural habitat. The series also includes guides to birds, and trees and shrubs.

The Concise British Flora in Colour by W. Keble Martin, 3rd edition, Ebury Press, 1974.

. . . to help you find out more about how and where they live

Useful background reading, providing information about particular groups of living things and about their habitats.

The Seashore and its Wildlife by Robert Burton, Orbis, 1977.

The Seashore by C. M. Yonge, Collins, 1976.

Underwater Life by Peter Parks and Oxford Scientific Films, Hamlyn, 1977.

The World You Never See: Insect Life by Theodore Rowland Entwhistle and Oxford Scientific Films, Hamlyn, 1977.

Nature's Hidden World: Woods and Forests by Michel Cuisin, edited by Michael Chinery, Ward Lock Kingfisher Library, 1979. Other books in this series deal with birds of prey, and lakes and rivers.

AA Book of the British Countryside, Drive Publications, revised edition, 1981.

The Natural History of Britain and Ireland by Heather Angel *et al.*, George Rainbird, 1981.

The Natural History of Britain and Northern Europe is a series produced by George Rainbird, dealing with the ecology and inhabitants of different habitats, including coasts and estuaries, and fields and lowlands.

Usborne Regional Guides describe the great variety of wildlife and the different habitats to be found in different parts of Britain; each book includes an identification guide.

Look for hidden plants and animals

Look for tracks and signs
Listen for sounds

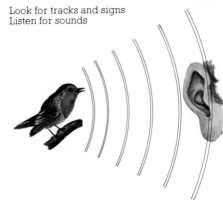

. . . to help you understand more about how ecosystems work

These books explain the basic principles of ecology, using many examples. They will tell you more about the ideas introduced in this book.

Food Chains by Jan Ethelberg, Adam and Charles Black, 1976.

Nature's Network by Keith Reid, Aldous, 1969.

Ecology by Peter Farb and the Editors of *Life*, Time-Life International, 1965.

. . . and how they change with the seasons

The Countryside in Spring by E. A. Ellis, Jarrold, 1975. There is a book for each season of the year.

Nature through the seasons by Richard Adam and Max Hooper, Penguin, 1977.

Winter Biology by Owen Bishop, Cambridge University Press, 1981.

Find out more about the plants and animals you discover by looking at them more closely and reading about them.

. . . to help you plan your own fieldwork

Family Naturalist by Michael Chinery, Roxby Press, 1978.

The Young Ecologist by Neil Arnold, Ward Lock, 1981.

Investigations in Woodland Ecology by T. Prime, Heinemann Education, 1970.

Seashore and Sand Dunes by S. M. Evans and J. M. Hardy, Heinemann Education, 1970.

A handbook for naturalists by Mark R. D. Seaward, Constable, 1981. A practical introduction to the study of natural history for 'beginners of all ages'.

Usborne Natural Trails are a series of practical handbooks for children on topics such as bird watching, insect watching, ponds and streams, seashore life, trees and leaves, wild flowers, wild animals, and woodlands. The first six of these are collected together in an *Omnibus* edition.

. . . to help you study ecology in more detail

These textbooks explain many of the basic principles of ecology – for people who have some knowledge of biology.

Fundamentals of Ecology by Eugene P. Odum, 3rd edition, W. B. Saunders, 1971.

Ecology by Robert E. Ricklefs, Nelson, 1973.

Ecology by Charles J. Krebs, Harper & Row, 1972.

The Economy of Nature by Robert E. Ricklefs, Chiron Press, 1976.

Energy Flow Through Ecosystems (Units 1–5, Block A, S323 Ecology), M. E. Varley *et al.*, Open University Press, 1974.

The Biosphere, W. H. Freeman, 1970. A collection of articles from *Scientific American*.

Animal Ecology by Charles S. Elton, Sidgwick & Jackson, 1966.

The Pattern of Animal Communities by Charles S. Elton, Methuen, 1970.

Studies in Biology, published by Edward Arnold, are a series of slim paperbacks written by experts and presenting up-to-date accounts of particular biological topics. Ecological topics include 1. *Ecological energetics*, 2. *Life in the soil*, 6. *Microecology*, 19. *Estuarine biology*, 51. *Ecology of small mammals*, 74. *Decomposition*, 91. *Boreal ecology*, 122. *Ecology of streams and rivers*, 139. *Ecology of rocky shores*.

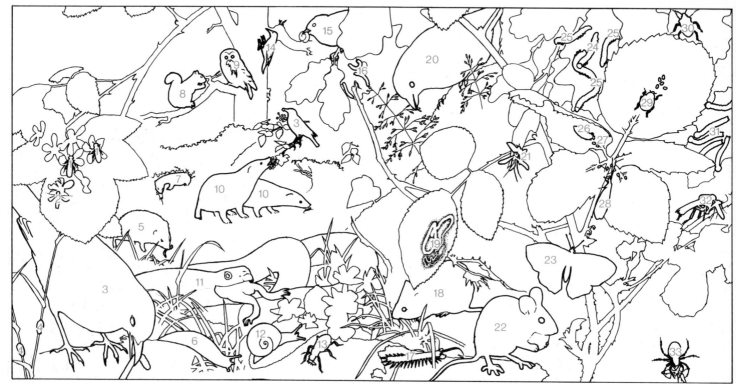

Key to the animals shown on pages 32 and 33

Herbivore	1	Honey bee *Apis mellifera*	Sucks nectar
Herbivore	2	A hoverfly Family Syrphidae	Sucks nectar
Omnivore	3	Blackbird *Turdus merula*	Eats mainly worms but also soft fruits
Herbivore	4	A froghopper *Cercopis vulnerata*	Sucks plant juices
Carnivore	5	Hedgehog *Erinaceus europaeus*	Eats snails, slugs, worms and insects
Herbivore	6	Great grey slug *Limax maximus*	Feeds on leaves
Carnivore	7	Weasel *Mustela nivalis*	Eats frogs, voles, rats and mice
Herbivore	8	Grey squirrel *Sciurus carolinensis*	Eats nuts, seeds and soft fruits
Carnivore	9	Tawny owl *Strix aluco*	Eats mice, voles, birds and young rabbits
Omnivore	10	Badger *Meles meles*	Eats small animals, seeds and soft fruits
Carnivore	11	Grass snake *Natrix natrix*	Eats small animals such as mice and frogs
Herbivore	12	Garden snail *Cepaea hortensis*	Feeds on leaves
Carnivore	13	Violet ground beetle *Carabus violaceus*	Feeds on small insects and mites
Carnivore	14	Green woodpecker *Picus viridis*	Eats bark insects and ants
Omnivore	15	Great tit *Parus major*	Eats mainly seeds but also insects
Herbivore	16	Winter moth caterpillar *Operophtera brumata*	Eats oak leaves
Carnivore	17	Centipede *Lithobius forficatus*	Eats small insects and mites
Herbivore	18	Bank vole *Clethrionomys glareolus*	Eats leaves, roots, shoots and seeds
Herbivore	19	A leaf-miner *Stigmella aurella*	Tunnels into leaves
Herbivore	20	Chaffinch *Fringilla coelebs*	Eats mainly seeds
Carnivore	21	A dance fly Family Empididae	Feeds on small flying insects
Herbivore	22	Woodmouse *Apodemus sylvaticus*	Eats nuts, seeds and soft fruits
Herbivore	23	Speckled wood butterfly *Pararge aegeria*	Sucks nectar
Omnivore	24	Dun-bar caterpillar *Cosmia trapezina*	Feeds on oak leaves and caterpillars
Herbivore	25	Clouded drab caterpillar *Orthosia incerta*	Feeds on oak leaves
Carnivore	26	Ladybird grub *Coccinella septempunctata*	Feeds on aphids
Herbivore	27	Oak leaf aphid *Thelaxes dryophila*	Sucks plant juices
Carnivore	28	Lacewing *Chrysopa septempunctata*	Feeds on aphids
Carnivore	29	Ladybird *Coccinella septempunctata*	Feeds on aphids
Herbivore	30	A shield bug *Pentatoma rufipes*	Sucks plant juices
Herbivore	31	Mottled umber moth caterpillar *Erannis defoliaria*	Feeds on oak leaves
Carnivore	32	A robber fly Family Asilidae	Feeds on other flies
Carnivore	33	Cross spider *Araneus diadematus*	Traps insects in its web

Key to the animals shown on pages 36 and 37

Herbivore	1	Grey coat-of-mail shell *Lepidochitona cinereus*	Feeds on small seaweeds
Herbivore	2	Bristled coat-of-mail shell *Acanthochitona crinitus*	Feeds on small seaweeds
Herbivore	3	Edible periwinkle *Littorina littorea*	Feeds on small seaweeds
Carnivore	4	Shanny *Lipophrys pholis*	Eats small animals such as shrimps, barnacles and crabs
Herbivore	5	Painted topshell *Calliostoma zizyphinum*	Feeds on small seaweeds
Omnivore	6	Common shrimp *Crangon crangon*	Filters particles from the water
Omnivore	7	Breadcrumb sponge *Halichondria panicea*	Sieves particles from the water
Carnivore	8	Common prawn *Palaemon serratus*	Feeds on worms and other small slow-moving animals
Omnivore	9	Herring gull *Larus argentatus*	Eats small animals and also scavenges for dead material
Carnivore	10	Ballan wrasse *Labrus bergylta*	Eats mainly mussels
Herbivore	11	Purple topshell *Gibbula umbilicalis*	Feeds on small seaweeds
Herbivore	12	Edible sea urchin *Echinus esculentus*	Feeds on small seaweeds
Omnivore	13	Common mussel *Mytilus edulis*	Filters particles from the water
Carnivore	14	Oystercatcher *Haematopus ostralegus*	Feeds on shellfish such as mussels and limpets
Herbivore	15	Common limpet *Patella vulgata*	Feeds on small seaweeds
Omnivore	16	Acorn barnacle *Balanus balanoides*	Filters particles from the water
Carnivore	17	Common starfish *Asterias rubens*	Feeds mainly on mussels
Carnivore	18	Cormorant *Phalacrocorax carbo*	Feeds on fish
Carnivore	19	Turnstone *Arenaria interpres*	Feeds on mussels, limpets and other small animals
Carnivore	20	Short-spined sea scorpion *Myoxocephalus scorpius*	Feeds on small fish
Omnivore	21	Peacock fanworm *Sabella penicillus*	Filters particles from the water
Carnivore	22	Beadlet sea anemone *Actinia equina*	Feeds on shrimps and other small animals
Carnivore	23	Common grey sea slug *Aeolidia papillosa*	Feeds on sea anemones
Carnivore	24	Common dogwhelk *Nucella lapillus*	Feeds on barnacles and mussels
Carnivore	25	Arctic cowrie *Trivia arctica*	Feeds on breadcrumb sponges and sea squirts

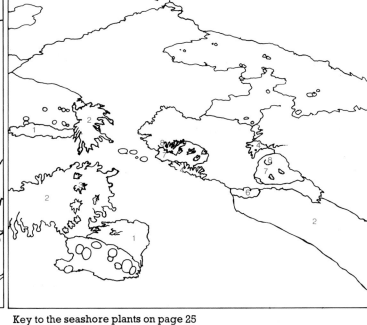

Key to the woodland plants on page 24

1 Bluebell *Endymion non-scriptus*
2 Bramble (or blackberry) *Rubus fruticosus*
3 Pendulous sedge *Carex pendula*
4 Holly *Ilex aquifolium*
5 Hornbeam *Carpinus betulus*
6 Lords-and-ladies *Arum maculatum*
7 Dog's mercury *Mercurialis perennis*
8 Pedunculate oak *Quercus robur*
9 Hazel *Corylus avellana*

Key to the seashore plants on page 25

1 Pepper dulse *Laurencia pinnatifida*
2 Knotted wrack *Ascophyllum nodosum*
3 *Polysiphonia lanosa*
4 Toothed wrack *Fucus serratus*
5 *Laminaria* species
6 Green laver (or sea lettuce) *Ulva lactuca*
7 Red calcareous alga *Corallina officinalis*
8 *Cladophora* species

You can only see the middle part of the shore in
this photograph. There are many other seaweeds
on different parts of the shore.
We have only marked some of the areas where
each seaweed is growing. These drawings will help
you pick out the different seaweeds.

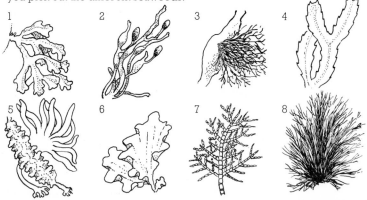

Names of animals shown in the book

Names of plants shown in the book

Woodland plants

bluebell *Endymion non-scriptus* 8, 24, 56, 75
bracken *Pteridium aquilinum* 8, 32
bramble (or blackberry) *Rubus fruticosus* 8, 24
    32, 33, 41, 73, 75
celandine, lesser *Ranunculus ficaria* 8
cow parsley *Anthriscus sylvestris* 66
foxglove *Digitalis purpurea* 8, 41, 73
hazel *Corylus avellana* 24, 66, 75
holly *Ilex aquifolium* 24
honeysuckle *Lonicera periclymenum* 8
hornbeam *Carpinus betulus* 24, 75
ivy *Hedera helix* 8, 32, 33, 56
lords-and-ladies *Arum maculatum* 24
mercury, dog's *Mercurialis perennis* 8, 24, 56
moss, silky fork *Dicranella heteromalla* 75
moss, white fork *Leucobryum glaucum* 56
oak, pedunculate *Quercus robur* 24, 28, 32, 33
    41, 56, 66, 68, 73, 75
primrose *Primula vulgaris* 8, 32, 33, 41, 53, 56, 71
sedge, pendulous *Carex pendula* 24, 75
soft grass, creeping *Holcus mollis* 8, 32, 33, 53, 56, 71
violet, common dog *Viola riviniana* 56

Seashore plants

Plants with no common name are listed under their
scientific name.
alga, a red calcareous *Corallina mediterranea* 8
    a red calcareous *Corallina officinalis* 25, 36, 37
Cladophora rupestris 8
Cladophora species 25
dulse, the *Palmaria palmata* 36, 37
    pepper *Laurencia pinnatifida* 25
Irish moss *Chondrus crispus* 8, 36, 37
Laminaria species 25, 36, 37
laver, green (or sea lettuce) *Ulva lactuca* 8, 25
    36, 37, 41, 68
phytoplankton 9, 31, 42, 62
Polysiphonia lanosa 25
wrack, bladder *Fucus vesiculosus* 36, 37, 40, 41
    knotted *Ascophyllum nodosum* 25
    toothed *Fucus serratus* 8, 25, 36, 37, 38

Answers to questions

Decomposer food chains   Page 43
These are the two food chains shown
in the pictures.

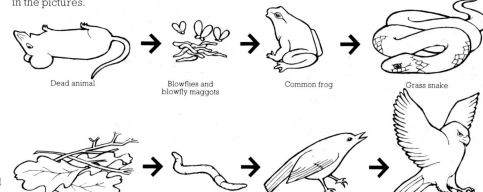

Dead animal → Blowflies and blowfly maggots → Common frog → Grass snake

Leaf litter → Earthworm → Blackbird → Sparrowhawk

Trophic levels   Page 57
In the food web on page 53 there are
four animals which are at trophic
level 2 – the earthworm, the snail, the
bank vole and the rabbit. They are at
trophic level 2 because they are
feeding on plants.

The tawny owl is at trophic level 3
because its energy has been
transferred three times:
• from the sun to plants
• from plants to the bank vole
• from the bank vole to the tawny owl.

The shape of a pyramid   Page 65
The upper pyramid has the more
efficient herbivores.

What would happen . . . ?   Page 74
The number of badgers in the area
would increase because there would
be more worms for them to eat.

Index